NATURE'S

CONSTANTS

π

2520

e

1260

Robert M. Oman, Ph. D.

Disclaimer and Terms of Use Agreement

This book is offered solely for educational and entertainment purposes. The research presented here should not be interpreted as predicting the time or place of any future event. The author and publisher make no representation or warranties with respect to the accuracy, applicability, fitness, or completeness of the contents of this book. The author and publisher accept no liability for any errors, omissions, misuse or misunderstandings on the part of any person who reads this book. The author and publisher accept no responsibility for any damage, injury or loss occasioned to any person as a result of relying on any material included, omitted or implied in the book. References are provided for informational purposes only and do not constitute endorsement of any other sources.

No part of this book may be reproduced or transferred in any form or by any means, graphic, electronic, or mechanical, including photocopying, or by any information storage retrieval system, without the written permission of the author.

Table of Contents

Preface

My early motivation for writing about constants was to help clarify in my own mind how the constants I have been using for many years in my scientific work came about, and how, and if, they were related to one other. After a short time my research led me to the result that many constants were the result of either simple mathematical manipulations or that they occurred in nature.

There are certain time intervals and possibly dates coded into nature. This study of nature's constants should not be construed as a book on prophecy or one based on "newly discovered" ancient manuscripts that showed the "end of the world" on a certain date. While such conclusions may be tempting for some, I have not found that the study of natural constants has any predictive value; possibly mild speculative value, but certainly not any definitive predictive value. Newton warned about such speculations, a warning I heartily agree with.

The study of natural constants contains many surprises. I may not have seen all of them, but I invite you to join with me on this fascinating journey. I originally thought this would be a rather boring exercise in putting together the pieces of a puzzle leading to known natural constants. It was not!

As distinguished a scholar as Isaac Newton spent many years of his life studying numbers and their relation to time (dates) and distance. He was a keen student of history and the relation of historic events to prophetic writings. His work in this area has only become publically available within the last couple of decades. It turns out that Newton spent more effort in trying to uncover basic knowledge of nature than physics and mathematics. More and more of his writing are being released by the THE NEWTON PROJECT at the University of Sussex. It is fascinating reading from one of the clearest expositors you will read, this despite the writing style and spellings of the 1600s. Newton was convinced that the dimensions of the temple described in the Book of Revelation contained information on the timing of significant future events. The answer may be in the future.

I discovered that the study of Nature's constants led me to think in areas I never anticipated. The outline of this study and the order of topics in the book are as follows:

1) There are constants inherent in nature.
2) Some of these constants lead to numbers related to distances.
3) And most fascinating, what could be called "derived," constants lead to time intervals and dates in history.

Whether the "predictions" are real or coincidental I leave to you the reader and future students of these relationships to discover.

I. Introduction

There are a few basic guidelines that need to be set out at the beginning of any discussion of the constants inherent in nature.

First, the Creator of the universe appears to have created an orderly universe in the respect that things like the reaction that produces heat in the sun, the orbiting planets with their various moons and the eco system of this earth all run quite well. The earth does not get out of orbit every once in a while. In fact many of the systems have built-in self correcting mechanisms.

Look at a simple system such as the position of a copper atom in a block of copper. This copper atom behaves as an ion core as do all the other atoms. (An ion core is the basic atom with one electron removed thereby giving it a positive charge.) All the atoms (ion cores) are in continuous motion about some equilibrium position, and they are held in place by the repulsive forces between the ion cores. (Remember that like charges repel.) So while all the atoms are in continuous motion, they all move around an equilibrium position. So despite the continuous motion of all the atoms in the block of copper, each ion core on average spends all of its time in the exact position where it should be, equally spaced between all its neighboring ion cores.

Many electrical systems are designed to be self correcting. A voltage monitoring device can be made so that if there is an increase in input voltage, the system compensates to maintain a set output voltage; and if there is a decrease in voltage, then the system compensates to again maintain a set output voltage. The whole universe seems to run like that. Ingenious, no?

If you accept the concept of a Creator, and if that Creator wanted to communicate with us, wouldn't you expect that communication would be in a form something similar to how the universe is constructed? Many scholars have studied the question of whether past and future historical events can be correlated to distance between places and the time between certain historical events. Most of these scholarly studies start with the early Jewish writers. It is easy to say that the Creator chose a people to be his special people on the earth and conveyed to them certain information. Whether you believe that or not the argument is supported by the simple fact that the Jewish people have maintained their integrity over a long period of time and with the same language and culture. Newton believed, as did other contemporaries, that the information in what is now referred to as the Old Testament contained information on the time of future events and the distance to important places where these future events would take place.

While a study of the history of places and future events did not fit with the original intent of studying the constants in nature, a small aspect of the work has turned in that direction. Any discussion of the constants in nature must start with numbers, the title of the next chapter. These numbers are then related to distance. And finally the constants in nature are related to distance and time.

II. Numbers

The constants that occur in nature fall roughly into three categories. First, there are constants that come from mathematics. These constants come from asking very basic questions of mathematical relationships. Next there are constants that come from often conceptually very simple scientific experiments. And finally there are the constants that come out of the sun-earth-moon system.

Just Plain Numbers

There are certain numbers that are interesting and have interesting features when operated on by certain mathematical operations. The phrase "operated on" is a phrase used by mathematicians to indicate that some specific thing (operation) is being done to a number. For example, when a number is multiplied by itself several times, that is a specific mathematical operation.

Start with the number 7. In the mathematical operation of multiplying 7 times 6 times 5 all the way down to 1, another number is generated. This operation is called the factorial and is indicated with an explanation mark (!). The factorial of a number is that number multiplied sequentially by all the integers down to 1.

$$7! = 7 \times 6 \times 5 \times 4 \times 3 \times 2 \times 1 = 5040$$

The number 5040 is an interesting number. Plato was quite a devotee of this number suggesting, among other things, that communities be organized of 5040 families.

The number 5040 can be divided cleanly by every integer from 1 to 10 and 12. In fact 5040 can be divided cleanly twice by 12. Clean division means that the result of the division is an integer, that is, it does not contain a decimal.

The number 5040 is also the product of $7 \times 8 \times 9 \times 10 = 5040$. Multiplication by 11, the first integer that does not divide 5040

9

cleanly, produces the number 7920.   The number 5040 and its divisions by 10 and 100 and by 2 and 2 again occur in studies of the constants in nature.

One-half of 5040 is 2520 which is the number written on the wall of Belshazzar's palace on the eve of the fall of Babylon brought about by the Medes and Presians[1] who had entered the city in the dry river bed after having diverted the river.

One-half of 2520 is 1260, the diameter in feet of the Avebury Circle in England, a very old monument.

One-hundredth of 2520 is 25.20 the number of inches in a cubit. There are many definitions of the length of the cubit, but all are equal to or close to 25.2 inches.

And a real interesting point about 7 and the product of 360 and 7 is the fact that 7 is the only cardinal number (Cardinal numbers are the integers 1 through 10) that does not divide cleanly into 360.   Check out the chart below.

360	Divided by	Result
360	1	360
360	2	180
360	3	120
360	4	90
360	5	72
360	6	60
360	7	51.42871
360	8	45
360	9	40
360	10	36

But the number 2520 is the product of 360 and 7. The 360 number comes from the number of days in the ancient year, that is, before the great flood. (Immanuel Velikovsky, *Ages in Chaos,* Sidgwick & Jackson, London, 1968) It is also the number of years in several

10

ancient calendars, particularly the Babylonian calendar. Note that the Babylonian number system was based on 60. The number 360 has been used for so long that it is accepted as an important number. The use of 360 as the number of degrees in a circle may have been influenced by this 2520 divided by 7 calculation. Further, John Dee, tutor to Queen Elizabeth (1558-1603), may have used the number 2520 in establishing the statute mile. There is more on this subject in a later chapter.

The Natural Constants

There are two very important natural constants, pi (π) and e. You have no doubt heard of π, the ratio of circumference to diameter for a circle. If you studied mathematics, you have also at least heard of e. There are other natural constants but these are the main ones.

Young students just starting out in schooling often determine π experimentally. Wind a string around a circular object such as a jar cover or cylindrical can. Measure the length of this circumference, then place the measuring instrument, usually a wooden ruler across the diameter and measure the longest distance across the circle. This is the diameter. A class of students doing this exercise will produce several ratios of circumference to diameter with average usually being very close to the accepted value of $\pi = 3.1416...$ The instruments used usually will not give four place accuracy, but with some machine tools such as micrometers and vernier calipers, typical measurement tools in secondary and college physics laboratories, π can be found to two decimal places with no problem.

The number π is what is called in mathematics a transcendental number. This is a mathematical word that means π cannot be represented by a fraction. There is a more formal definition but this one is sufficient. The mathematical proof that π is a transcendental number was worked out in 1882, but for many years mathematicians continued to search for the fraction that would produce the number π. Early in the 20th century that quest was abandoned, and by mid-century (circa 1950) an early computer was made to run long enough to calculate the value of π using an accepted algorithm to so many decimal places that the question of finding a fraction for π was totally abandoned. The number π is a unique number that comes out of a very simple mathematical question and physical experiment: "What is the ratio of circumference to diameter in a circle?"

Another transcendental number that comes from straight number manipulation is the mathematical constant e. This number is the product of a simple mathematical relationship.

$$e = \left(1 + \frac{1}{n}\right)^n$$

The number e is generated using this formula to make successive approximations, that is, evaluating the formula for higher and higher values of n. While this formula may look formidable, when you start making the calculations it is not that difficult. Take n as 1. Inside the parenthesis the number is 1 plus 1 over 1 or 2 for the parenthesis. The n in the superscript position just tells how many times to multiply the parenthesis by itself. This is called "raising to a power," that is, successively multiplying what is in the parenthesis to the prescribed power, n. The number 2 then, when raised to the power 1, is just 2. Now make the n equal to 2 so inside the inside of the parenthesis is 1.5 but this 1.5 is multiplied by itself 2 times to give 2.25. With a hand calculator you can make better and better approximations by picking larger and larger numbers for n as shown in the table below.

n	$1 + \dfrac{1}{n}$	$e = \left(1 + \dfrac{1}{n}\right)^n$
1	2	2
2	1.5	2.25
10	1.1	2.59
100	1.01	2.70
1000	1.001	2.72

12

Successive calculations approach the number for e of 2.7183... This definition allows calculation of e to whatever desired accuracy you like. With hand held calculators 7 decimal places is easily calculated.

The number e, similar to π, is a transcendental number (Remember: transcendental numbers cannot be written as fractions.). But this is not the end of the story of e.

Another way to look at e is with a graph. If a trend is showing a rapid rise, we often say the rise or increase is exponential. This refers to a somewhat unusual algebraic relationship. In the world of algebra a direct relationship is one of the forms such as the cost of commuting is directly related to the commuting distance. That relation is expressed symbolically as $\$ = kD$, where the cost is represented by the $\$$, k is some constant that ties the two things together and D is the commuting distance. These kinds of relations work well.

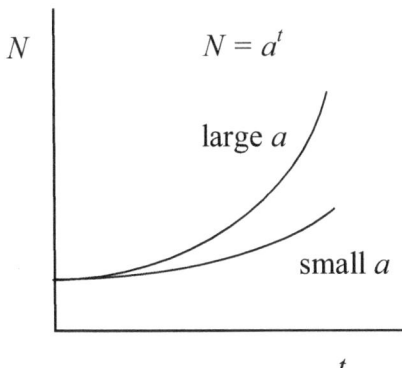

However when it is appropriate to describe a rapidly rising situation, the relationship for say the growth of fruit flies surrounding rapidly ripening fruit might look something like $N = k(a^t)$ where N represents the number of fruit flies at a time t, with k again being a constant that relates the variables. The number a can be chosen to fit the situation or more to the point the actual data. If the "a" in the exponent relation is large, then the number of fruit flies grows very rapidly, while if "*a*" is small the number of fruit flies grows slowly over time. The rate of growth at any time *t* is the slope of the graph at that time.

This, what is called an exponential relationship, is a curve of the general shape shown in the graphic. The slope of this curve at any point is a tangent line parallel to the curve. Imagine a straight edge being moved along the curve and always tangent to it where the straight edge is the slope of the curve at whatever point it touches.

Now for the really interesting question: Is there a number for the "*a*" value such that for any value of *t*, the corresponding number *N* and the slope are the same? Yes, there is, and by now you have suspected that that number is *e*. So by asking a very simple question about exponential relations the number *e* is again generated. Detailed calculations verifying this fact are given in text books.[2]

Our third and final look at this mysterious constant starts off in a direction that seems about as far as you can get from generating a number. This is an excellent mystery with "our hero," the number *e*, appearing to save the day so to speak at the end. The story starts with a discussion of logarithms. Before you stop reading, I have to tell you that these are not the logarithms you might have studied in secondary school. No manipulations or logarithm type calculations are involved. Just enjoy the beauty of the mathematics without getting involved in the detail. There is only a little detail involved, and it will be done for you.

After getting you all convinced that this is going to be easy, if not fun, I am going to define logarithms in a way that, unless you studied mathematics at the university level, you have not seen. The formal definition is

$$\ln x = \int_1^x \frac{dt}{t}$$

This clearly needs explanation. The $\ln x$ is clear enough; it is the logarithm of some number which for the moment is designated by the symbol x. (The ln is used rather than the traditional log because this definition is for the natural logarithm.) That long "S" shaped symbol means to take a summation and the subscript and superscript define the region for the summation. OK, so we are supposed to sum something from 1 to x. And the thing to sum is an area defined by the area under the curve of $1/t$. The graphic shows the curve and the area. The sum for the area shown is in the t direction as indicated by the t-axis.

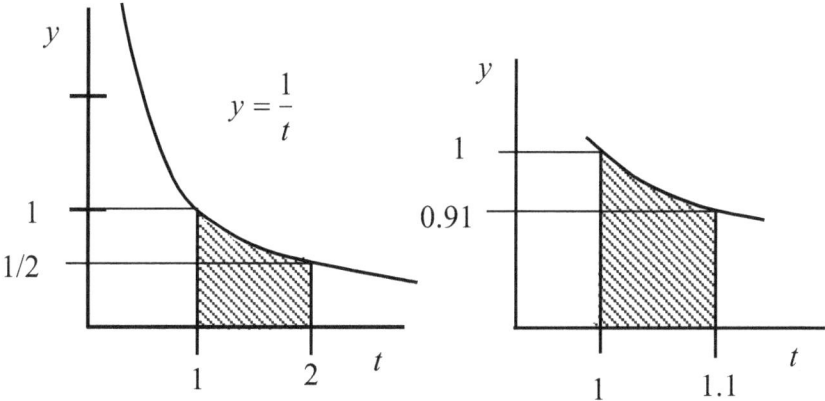

The graphic shows the curve $y = 1/t$ as a solid line. Understanding the shape of this curve is not so bad if you keep the numbers simple. The place to start is at $t = 1$. When $t = 1$, $y = 1$. When $t = 2$, $y = 1/2$ and when $t = 3$, $y = 1/3$ and so on for larger and larger t values. As t increases, the value of y drops toward zero. For t values less than one, the y increases, but for this exercise that part of the curve is uninteresting. The definition of the ln defines the area starting at $t = 1$. In this graphic the area shown represents the natural logarithm of 2. Yes, that is correct; the natural logarithm is an area. Specifically it is the area under this curve out to the value associated with the desired logarithm. The ln 3 is the area under this curve from 1 to 3. The ln 6 is the area under this curve from 1 to 6.

To see how the number e comes into play, calculate the ln of the number 1.1. Look at the expanded graphic that shows a small piece

15

of this curve from 1 to 1.1. Consider the piece of the curve between $t = 1$ and $t = 1.1$ as shown. The area under this part of the curve is approximated by the area of a rectangle $0.10 \times 0.91 = 0.091$, and the area of the small triangle sitting atop this rectangle calculated as one-half the base times the height: $(1/2)(0.10)(0.09) = 0.0045$.

The total area of this rectangle and triangle is 0.0955, thus $\ln 1.1 = 0.0955$ If you have a hand calculator, take the ln of 1.1. Most hand calculators will give the natural logarithm of 1.1 as 0.0953, which is just a little bit smaller than this number as is expected because, as can be seen in the graphic, the actual area is a little less than the rectangle and triangle used in this approximation.

The connection between this definition of the natural logarithm and the constant e is amazing! The constant e raised to the power of the area under the curve is equal to the upper limit of the integral. Don't worry about raising e to some decimal power. That inexpensive few dollar calculator will do it for you or just check below.

$e^{area} =$ upper limit of area calculation

And remember that this area is the ln of the number that is the upper limit of that summation.

$\ln($ the upper limit of the area calculation$) = area$

Verify for yourself that $e^{0.0953} = 1.1$, and that $\ln 1.1 = 0.00953$. And we already calculated this area as 0.00955.

Again, a reasonably simple area problem in calculus produces the number e, which occurs other places in nature.

With establishment of e as a number that can come from one of either of the definitions discussed, it is now time to take another excursion into a seemingly unrelated subject.

When Pierre and Marie Curie first discovered radioactivity, the decay of certain materials by the emission of certain unique particles, they defined this radioactivity in terms of what they called the activity of the sample. If you consider a sample of radium, as they did, it is impossible to predict when each radium atom will give off some radiation, that is, emit some particle. However it is not impossible to define an activity, the number of decay products per unit of time for a large collection of radium atoms. The activity of any collection of decaying material is measured by the number of decay products per second. The historically first unit of activity was the Curie (Ci) which is equivalent to 3.7×10^{10} decays per second. The Curies chose this number because it was the activity of 1 gram of a readily available radioactive material, radium.

So, while it is not possible to predict when a particular radioactive nucleus will decay, it is possible to predict the decay rate, that is, the activity, of a large collection of nuclei. This rate depends on what nucleus is decaying and the number of decaying nuclei in the sample. Simply put, the more material, the more decays. In equation form the activity or decay rate is written as

$$A = \frac{\Delta N}{\Delta t} = \lambda N$$

Again, the explanation of this mathematical statement is fairly simple. The activity, A, is the number of decays, ΔN, per or divided by an increment of time, Δt. And this decay rate is proportional to N, the number of nuclei in the sample and a constant, λ, which varies with the material. The proportionality constant λ is positive for growth and negative for decay. Said another way, the number of particles emitted from a sample of radioactive material is proportional to the amount of material and some constant associated with that material.

Calculus people use the script symbol d to represent a change in something and would write this relation as

$$\frac{dN}{dt} = \lambda N$$

With a little algebra this statement can be rewritten as

$$\frac{dN}{N} = \lambda t$$

The dN/N is in the same form as the definition of the natural logarithm, and remember the natural logarithm involves e raised to a power. Leaving out a few calculus steps the solution is

$$N = N_o e^{\lambda t}$$

where the N_o is the initial amount of material when time started to be counted. This statement gives us the amount of material (remaining), N, at a time t starting with N_o for a material with a particular constant λ.

A similar experiment is often performed in elementary or secondary schools. In this classic experiment a certain mass of yeast is placed in a water environment conducive to growth. After some time, usually over a weekend, the culture of yeast is dried and again massed. The mass is proportional to the number of growing yeast, and the increase in mass is the same as the increase in N. Knowing the initial and final mass and the time, the λ. can be determined.

This growth and decay process following the

$$N = N_o e^{\lambda t}$$

occurs in a myriad of places in nature from learning curves, to growth of business, to physical cooling, many electrical circuits and to the clocks that are used to determine ages of certain materials on the earth.

18

The situation is simply stated as follows: Start with an amount of radioactive material containing N radioactive nuclei. And while we cannot predict when each nucleus will decay we can predict that for this N number of nuclei the rate of decay (number of decay products per second) is proportional to the total number of nuclei present.

After this intense discussion a little relief is in order in the form of this cartoon of the precocious little characters Newt and Leibe named after the inventors of Calculus, Newton and Leibniz. They express a simple truth about growth and decay.

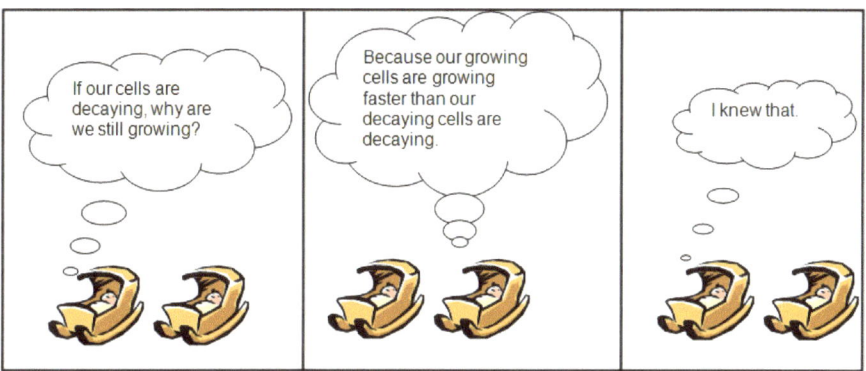

1. Daniel: 5 (25-30)

2. Oman and Oman *Calculus for the Utterly Confused,* second edition McGraw-Hill 2007

III. The Nautical Mile and the Metric System

The number that is most crucial to establishing the nautical mile is the number 360. Half of the 5040 (Remember 5040 = 7!) is 2520 which is also the product of 360 and 7. The number 360 also comes from ancient calendars, many of which are based on a 360 day year with twelve 30 day months. A very good discussion of the 360 day year and the possible mechanism that led to the present 365.25 day year is given by Immanuel Velikovsky in the book *Worlds in Collision*.[1] In *Worlds in Collission* Velikovsky discusses the early calendars and speculates that the change in the length of the year was a result of a massive visitor passing close by the earth altering the dynamics of the earth's rotation and its rotation about the sun. This visitor dumping ice/water on the earth accounts for the great flood that is recorded in ancient cultures, the apparent shift in the magnetic poles, and the large quick frozen mammals, some even with vegetation in their mouths, that are preserved frozen in northern latitudes. More on the possible mechanisms leading to a longer earth year are presented in Appendix A.

Longitude and Latitude

Rotation in the east to west direction on the earth is measured by 360 longitude lines with line zero, called the Prime Meridian. The prime meridian historically has been placed to pass through different places, usually cities associated with dominant maritime nations. At present, the Prime Meridian passes through Greenwich, England. These meridian lines all pass through the poles and they are spaced one 360th of the way around the earth at the equator or any latitude. The latitude lines are spaced at degree intervals up and down from the equator. There are 90 degrees from equator to North Pole and 90 degrees from equator to the South Pole. The graphic shows these lines on the earth.

The distance around the earth in nautical miles is 360 times 60 miles for each degree of latitude or 21,600 nautical miles. This 60 nautical miles per degree of latitude is approximately equivalent to 69 statute miles. The definition of the statute mile is taken up in the next chapter. The nautical mile is defined in terms of latitude because the distance between longitude lines is different depending on latitude. At the equator the distance between longitude lines is this 60 nautical miles but for increasing latitudes, that is going north or south, the distance between longitude lines grows shorter for increasing latitude. At 90 degrees north or south latitude, the longitude lines converge.

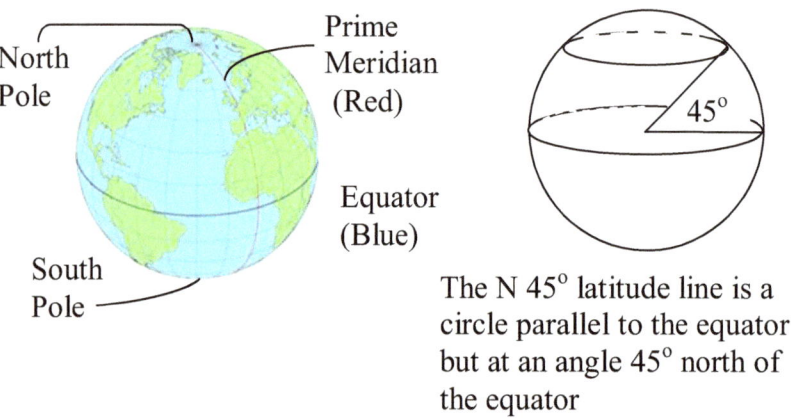

The N 45° latitude line is a circle parallel to the equator but at an angle 45° north of the equator

Latitude lines are parallel to the equator.
Longitude lines pass through the poles.

Navigation

Longitude and latitude can be determined by astronomical observations. These are very time-consuming and in the case of longitude require much skill and lengthy computations. While longitude and latitude of points on land, particularly harbors, can be established this way, establishing latitude and longitude this way is very difficult on a moving ship.

The length of the day varies from day to day throughout the year, but the position of the sun above the horizon at high noon is relatively easy to establish even working on the tossing deck of a ship. Knowing the latitude by measuring the angle of the sun at high noon for shipboard navigators was the first step in navigation.

Latitude, in the northern latitudes, was also established from the position of the North Star. High noon is defined as the highest angle of the sun above the horizon. For early navigators, sailing to a specific longitude and latitude consisted of sailing in the general direction of where you wanted to go using speed of the ship and magnetic compass direction, and intercepting the latitude for your desired destination. The ship was then sailed along this latitude until the destination was reached. If you choose the wrong direction to sail, then you would move away from your intended destination. This happened.

The earth does one complete rotation in 24 hours. Selecting the number of hours in a day has to take into account that the number be divided by 2, so the hours are somewhat evenly divided into day and night. At the equator, the day, as defined by sunrise and sunset, does divide the time very close to half daylight and half darkness. How many hours to assign for day and night is the next question. The number 12 (12 for daylight and 12 for darkness) is most convenient because 12 is easily divisible by 2, 3, 4 and 6. Sixty minutes per hour and sixty seconds per minute probably came from the Babylonian base sixty number system. Remember: The Babylonians had a 360 day calendar.

High noon at the prime meridian, which at the present time, and by international agreement, passes through Greenwich, England, is set by the maximum angle of the sun above the horizon and defines 12 noon, the mid-point of the day. The 360 degrees divided by 24 hours per day for one rotation means there are 15 degrees of rotation for each hour. As the sun is moving from east to west across the earth, high noon at Greenwich corresponds to 11:00 AM at 15 degrees, 10:00 AM at 30 degrees and so on around the globe. High noon of any day is set by high noon in Greenwich, but half way around the world at longitude $180°$ it is midnight and that day is just starting.

To determine longitude at sea, first establish high noon and then determine the time difference between your high noon and high noon at Greenwich. If your high noon occurs at Greenwich Time, called GMT, of 11:00 AM, then the appropriate longitude is 15 degrees or one hour east of Greenwich. (The sun travels from east to west around the globe.) If high noon occurs at 2 hours after noon or 1400 hours, then it took the sun 2 hours to travel 30 degrees to the west to this location, and the appropriate longitude is 30 degrees east. The accuracy of longitudinal location thus depends on an accurate clock that can maintain reasonable accuracy for ocean crossing voyages.

In the late 1600s there were hundreds of British ships traveling between Britain and the West Indies all navigating by speed and compass, and latitude. The problem of wrecks and lost time and treasure because of insufficient navigational information was so bad that in 1714 the British Parliament offered a prize of £20,000 for a solution to the longitude problem with an accuracy of one-half a degree, equal to 30 nautical miles, for a round trip from England to the Caribbean and back. Such a voyage took roughly six weeks and required an accuracy of better than 3 seconds a day. The first version of a seagoing clock was taken on a trial voyage in 1736. The clock was the work of a scientifically obscure but experienced and well- respected clockmaker, John Harrison. After producing three large clocks, all successful, Harrison in 1759 produced a clock that looked more like an enlarged pocket watch than the large clocks of an earlier era. In 1760 the Harrison "watch" started off on the England to Caribbean and return trial. Before very long, however, a problem developed requiring a stop for replacement of lost food supplies. The Harrison clock was so accurate in locating the resupply port that the Captain of the ship immediately offered to purchase the first clock the Harrisons (father and son were now working on the clocks) offered for public sale. Despite the clock doing everything required to receive the prize, payment was withheld due to jealousy and bureaucratic bungling. Half the original prize money was awarded Harrison in 1773 after many more trials and further restrictions added to the original specification of accuracy. Despite the efforts of English bureaucracy at delaying the acceptance of the timepiece, the watch and its many successors and imitators did solve the longitudinal problem. By 1850 accurate clocks were in general use in ocean going sailing ships. This whole

24

twisted story of longitude is in the fascinating book *Longitude* by Dava Sobel.[2]

The Metric System

The metric system is based on standards for length, mass and time. First, look at time. A day is established by the sunrise on successive mornings. While the length of day and night varies throughout the year, the variation between the lengths of days is slight. The second was established initially as a fraction of the mean (average) solar day. The day is divided into 24 hours, with each hour further divided into 60 minutes and each hour again divided into 60 seconds.

In 1791 the meter was initially set by the French Academy of Sciences as one ten-millionth of the distance from the North Pole to the equator along a longitude line passing through Paris. This distance was based on astronomical observations. What is not generally known is that the meter can be directly related to a swinging pendulum. A common elementary physics experiment involves a pendulum where the experimenter is asked to vary the angle of the arc swept out by the oscillations, the mass of the swinging bob and the length of the pendulum. Simple experiment shows that for small angles of less than 4 to 10 degrees maximum, the time for one complete round trip of the pendulum, the angle and the mass of the bob have no bearing on this time. The time for one half of a complete round trip excursion, which is from one extreme position of the motion to the other, is one second for a pendulum length of one meter. Galileo was one of the first to look into the oscillations of a pendulum. He noticed that in the church he attended each morning as a student at Pisa the candle chandelier was lit by pulling the chandelier to one side of the church where the priest then lit the candles and let the chandelier oscillate. Galileo noted that the time of one excursion did not change as the angle decreased over time. He timed the oscillation with his pulse. Legend has it that Galileo developed a pendulum of approximately one meter length and adjusted the length of the pendulum to establish a resting heartbeat for each of a physician's patients. When the heartbeat increased, as measured by matching a current reading with the resting one established for that patient, the physician could definitely determine the amount of increase in heart rate of the patient. What is most interesting is that the heart rate of a healthy person corresponds to one excursion (maximum to maximum) of a one meter long pendulum. The second (determined by dividing the day into 24 hours each hour into 60 minutes and each minute into 60 seconds) produces a length of time for one complete excursion of a pendulum that is equivalent to one ten-millionth of the length along a longitude from pole to equator and the time between heart beats of a healthy person. The pendulum was proposed as the meter standard but abandoned because the oscillation time depends on the gravitation constant which varies over the surface of the earth, or more specifically the distance from the center of the earth.

The mass unit can also be tied to the meter. A cube of water one one-hundredth of a meter (called a centimeter) on a side is called a cubic centimeter. This cubic centimeter of water has a mass of one gram. A cube of water 10 cm on a side contains 1000 cubic centimeters and has a mass of 1000 gram or 1 kilogram. The amount of energy to raise one gram of water one centigrade degree is one calorie. (The more popular food calorie is 1000 calories or 1kCal as it is often written.) In this way thermal energy, and also electric energy which is the common way of heating water, is directly connected to the heat unit via length and mass units.

The meter and the second are closely tied together by the oscillations of the pendulum, and convenient mass units come out of certain volumes of water, the most abundant element on the planet.

1. *Worlds in Collision*, Immanuel Velikovsky, Simon and Shuster, New York, 1977.

2. *Longitude*, Dava Sobel, Penguin Books, New York, 1995.

IV. The Statute Mile, the Foot and the Inch

The origin of the statute mile dates back to Queen Elisabeth and her mathematics tutor John Dee. While the emergence of England as a major world power during the reign of Queen Elisabeth (1558 to 1603) is attributed by most historians to her political ability, her intellect and ability to find and use the leading intellectuals of her time was perhaps at least equally important in the ascendance of England to global power status. Navigation on the sea was critical for any country that wanted to be a global power. And John Dee understood navigation.

John Dee, born in 1527, was a leading mathematician and all round scholar. He was a contemporary of, and collaborator with Mercator, the geographer instrumental in producing a map on a piece of paper which when laid flat portrayed the geography of the earth in terms of longitude and latitude. Dee was also an occult mystic according to the 16th century definition of the phrase. While no record of the thinking concerning the logic of the statute mile exists, a reasonable guess as to the thinking process can be made.

Starting with the numbers 2520 and pi, multiplication produces roughly 7917. Take this number as the diameter of the earth in statute miles and multiply by pi again (pi times the diameter of a circle gives the circumference) to find the circumference of the earth as 24,871 statute miles. This method of calculating the circumference of the earth defines the statute mile. No one is going to wrap a string around the earth at the equator to get a string of length 24,871 statute miles, but one degree of longitude at the equator is 24,871/360 or 69.1 miles and one tenth of a degree is 6.91 miles.

The sun moves 360 degrees in 24 hours, or 15 degrees per hour. This corresponds to 1/4th of a degree per minute. One-fourth of a degree corresponds to 17.3 miles. There is an easier way to see this using a unit conversion equation. Start with 360° of rotation in 24 hours corresponding to 24,871 miles per 24 hours and multiply by 1 hour over 60 minutes and again by 1 minute over 60 seconds to get the rate of 0.29 miles per second.

$$\frac{360^o}{24 \text{ hr}} \Rightarrow \frac{24,871 \text{ mi}}{24 \text{ hr}} \left[\frac{1 \text{ hr}}{60 \text{ min}}\right] \left[\frac{1 \text{ min}}{60 \text{ sec}}\right] = \frac{0.29 \text{ mi}}{\text{sec}}$$

Experimentally a length standard can be made using this information. Go to the equator and make observations of high noon using the highest point of the sun above the horizon at two points along the equator. If the observations are 5 seconds apart then the observation points along the equator are separated by

$$\frac{0.29 \text{mi}}{\text{sec}}\left[5 \text{ sec}\right] = 1.44 \text{ mi}$$

While the mile is an appropriate size for navigation at sea, everyday interactions require smaller units. It would be convenient to have a length unit comparable to the size of hands or fingers, another one for the size of feet, another for the length of reach (tip of fingers to nose) and shoulder to tip of fingers.

If pi is multiplied by 7 five times the result is 52,800.

$$\pi \times 7 \times 7 \times 7 \times 7 \times 7 = 52,800$$

If the mile were divided by this number, the length would be comparable to hand size. However, the mile was divided by one-tenth of this number, 5280, to obtain a length comparable to the foot of a person. The foot length was further divided by 12 to obtain the inch. The number 12 is convenient because it is easily divisible by 2, 3, 4 and 6. The length from tip of fingers to nose comes out to about 3 feet, a convenient length designated as the yard, and the length from shoulder to tip of fingers is roughly 25 inches.

It is possible to speculate that if the mile were divided by 52,800, a length a little longer than the inch would have resulted, and multiplication by 10 would produce a length comparable to the present day foot. But multiplying by 10 is a problem because 10 is only cleanly divisible by 2 and 5. The advantage of 12 inches per foot is the ease of defining 1/2 or 1/3 or 1/4 of a foot.

The roughly 25 inches for the length from shoulder to tip of fingers is known as the cubit. Cubits have been used for centuries and it is tempting to assign 25.2 inches (one one-hundredth of the 2520 number) to the length of the cubit. The most famous use of the cubit is in the building of the Ark of the Covenant which is specified as a brick-like structure with square ends of 1.5 cubits and length 2.5 cubits

If we take one one-hundredth of the number 2520 that number is 25.20. The cubit as defined most popularly in the dimensions of the Ark of the Covenant is generally taken to be in the neighborhood of 25 inches. Newton investigated the use of the cubit as a measure throughout history. With the cubit defined this way (as 25.2 inches), the inch is approximately a thumb width and the foot, 12 inches, the length of a human foot. Why 12 inches? The number 12, besides being roughly the length of a human foot, is convenient because of the easy divisibility. Numbers (in inches) close to the length of a human foot, for example, 11 or 13 or even 10 or 14 do not have this convenience of division into halves, thirds and quarters.

The statute mile was established by Queen Elisabeth in 1592 based along the lines of the arguments presented here. Unfortunately, there are no physical records of the discussions between John Dee and Elisabeth concerning the mile, the foot and the inch. The argument presented here is reconstructed from educated guesses based on our understanding of their thinking on the subject.

While the nautical mile is convenient for global navigation, the statute mile and its subdivisions into feet and inches is more useful for everyday measurements.

V. Earth, Moon, Avebury Circle and Stonehenge

The numbers 7, π, 360 and 2520, the product of 360 and 7, all appear in the earth and moon dimensions. The earth has a diameter of 7 × 360 × π while the moon has diameter 6 × 360. And multiplying these diameters by π produces the circumference.

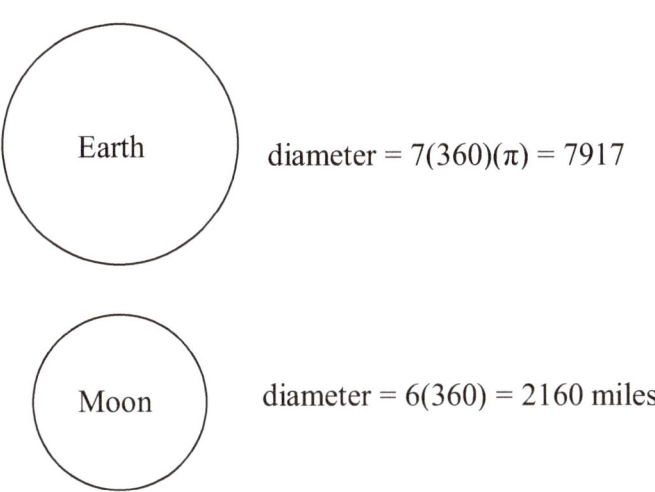

Earth   diameter = 7(360)(π) = 7917

Moon   diameter = 6(360) = 2160 miles

The diameter of the earth at the equator is larger than this number, while the diameter of a great circle through the poles is less than this number.

Eclipses

During an eclipse, when the moon comes between the sun and an observer at the right spot on the earth, the moon completely blocks out the sun. Viewed from the earth the sun and moon appear to be the same size. They are not, because the sun is clearly further away than the moon. If the Creator wanted to get the attention of people on earth, an eclipse, where the sun is completely covered by the moon, is certainly the way to do it.

Avebury Circle

The Avebury Circle in Wiltshire England is the largest structure of its type in Europe. Historians are unsure of the age of the structure though references to it in the writings of Julius Caesar indicate that it predates Roman occupation of England.

While erosion has taken its toll on the dimensions of the circle it appears to have an inner diameter of 1260 feet. If the inner diameter is 1260 feet and the width of the ditch 210 feet, which is quite possible considering the present width of the ditch and the length of time since original construction, then the circumference is a statute mile. The data producing this statute mile circumference is clearly vague, close but vague. Using $1260 + (2)(210) = 1680$ feet, this length in feet times π is 5278 feet, very close (within 1%) to a statute mile especially considering the obvious rough estimate of the diameter. It is interesting that the definition of the mile came centuries after the construction of Avebury Circle.

The original circle appears to have consisted of an outer circle of large stones, two inner circles made of smaller stones and two stone walkways across the circle.

What is truly amazing is that the structure is located at 51.4271 degrees of latitude. Recall from the table in Chapter III, Numbers that 51.4271 is equal to 360 divided by 7. And the number 7 is unique in that it is the only number between 1 and 10 that does not divide cleanly into 360. At 51.4271 degrees North latitude, Avebury Circle is located one-seventh of the distance from the equator to the North Pole along a longitude line.

Stonehenge

The structure at Stonehenge is the most popular structure of this type in the world. It is about 25 miles south of Avebury Circle. While the site is smaller and in better condition than Avebury Circle, there seems to be little in the way of numbers that can be gleaned from the site. Perhaps they are well hidden and will be revealed in time.

VI. Golden Ratio, Fibonacci and the Ark of the Covenant

There are a great many numbers and ratios that can be seen in the dimensions of humans and animals and in popular flowers and plants. This short discussion can only provide a glimpse of these numbers. To thoroughly cover the subject would take several volumes. The first ratio discussed has been used for centuries by artists and architects in studying the length relationships in the human body and the construction of structures that are pleasing to most humans.

The Golden Ratio

The most convenient definition of the golden ration is with line segments. Place two line segments, *a* and *b*, together in a straight line and make *a* longer than *b*. The golden ratio is defined as when the sum of both lengths divided by the longer length is equal to the longer length divided by the shorter length. This is easier to see in the formula.

$$\frac{a+b}{a} = \frac{a}{b}$$

The golden ratio is defined as the ratio of *a* to *b*, when the sum of the line segments divided by the largest segment is the same as the ratio of the larger segment divided by the shorter segment. With sufficient number manipulation the ratio is (approximately) 1.618, the golden ratio, which is an irrational number. That means that the number is a never ending decimal and cannot be expressed as the ratio of two numbers.

Try this simple experiment. Measure the length of your extended hand (wrist to tips of fingers), and the length from your elbow to wrist. The ratio of these lengths (elbow to wrist over hand) for most people is the golden ratio to accuracy consistent with the accuracy of your measurement.

The ratio is a pleasing dimension for paintings and architectural structure and has been used for centuries. Among man-made objects, dominant structures such as the Cathedral of Notre Dame in Paris and the Parthenon in Athens contain this ratio. The ratio is even found in the DNA molecule where the length of the spiral is 34 units and the width 21 units with ratio 1.619, very close to the golden ratio.

Golden Ratio from Simple Number Manipulation

The golden ratio can also be defined as the answer to a simple number manipulation question: Is there a number whose reciprocal is the decimal associated with that number? No need to perform any complicated math. Take the golden ratio as defined by the line segment ratio definition which is 1.61803399 . . . , and take the reciprocal using your hand calculator. The answer will be the decimal associated with the golden ratio or 0.61803399 . . .

Fibonacci Numbers

Leonardo of Pisa (1170 – 1240) aka Fibonacci is associated with a series of numbers called the Fibonacci series. The numbers are generated quite simply starting with the number 1. The first number is 1 and each succeeding number in the series is formed by adding the previous two numbers. Adding 1 and 0 produces another 1 and adding 1 and 1 produces 2. Adding 2 and 1 produces 3 and adding 3 and 2 produces 5 and so on. The Fibonacci sequence is the following:
1, 1, 2, 3, 5, 8, 13, 21, 34, 55, 89, 144, 233, 377, 610, 987, 1597, 2584, 4181, 6765, …

Pascal's Triangle

The triangular array shown here was popularized by Blaise Pascal (1623 – 1662). It is most convenient for finding the coefficients of the expression $(a + b)$ raised to some arbitrary power, not something everyone does in their spare time! But, a surprising property of the triangle is that it can also generate Fibonacci numbers.

The triangle is made starting with 1. The next row contains two numbers, the next three numbers and so on down the triangle. Each number is placed in the row mid-way between the two just above it and is the sum of those two numbers. The first row contains 1. The second row contains 1 and 1 placed as shown. The next row contains 1, 2 and 1. The next row contains 1, 3, 3 and 1. And the next row contains 1, 4, 6, 4, 1.

Adding the diagonals as indicated generates the Fibonacci numbers! This simple array generates Fibonacci numbers.

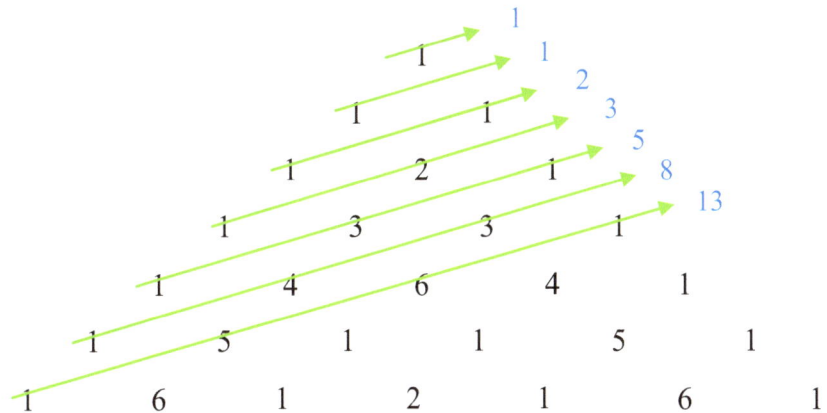

Flowers and Fibonacci

The evidence of Fibonacci numbers in nature is most easily seen in flowers. Most flowers have a number of petals equal to a Fibonacci number. Flowers with one, two and three petals are common. Flowers with four petals are rare. There are some flowers with 5, 8 and 13 petals but there are many flowers that have a number of petals equal to the larger Fibonacci numbers. One, three, eight and thirty-four petal flowers are very popular. The flowers shown here have one, three, five and eight petals with the common field daisy (far right) having 34 petals.

Fibonacci Ratios

There is another sequence that can be generated with Fibonacci numbers which has a remarkable result. If any Fibonacci number is divided by the following one, then another series of decimals is generated.

$$\frac{1}{1}=1 \quad \frac{1}{2}=0.5 \quad \frac{2}{3}=0.67 \quad \frac{3}{5}=0.6 \quad \frac{5}{8}=0.625 \quad \frac{8}{13}=0.615 \quad \frac{13}{21}=0.619$$

Now write the reciprocals of these numbers which is equivalent to each Fibonacci number divided by the preceding one.

1   2   1.5   1.67   1.6   1.625   1.615

If the calculation is carried out long enough it produces the same number as the golden ratio.

Noah's Ark and the Ark of the Covenant

The construction of Noah's Ark is specified in Genesis 6:15 to be 300 cubits in length, 50 cubits in breadth and 30 cubits in height. In this specification the breadth to height ratio is 5:3, both Fibonacci numbers and with ratio equal to 0.6, quite close to the golden ratio. As a practical matter one would not expect Noah to be able to construct the ark with dimensions of height to breadth ratio to several decimal places. When making something out of wood, this would have been an impossible task, especially in a time of high humidity!

The Ark of the Covenant is specified in Exodus 25:10 to be of length 2 and one-half cubits and the breadth and height equal to one and one-half cubit. Here the ratio of breadth or height to length is again 3:5.

What conclusions can be drawn from these facts from nature? First, the Creator of the natural system likes numbers and is partial to Fibonacci numbers. Second the Creator likes algebra, as the simplest definition of the golden ratio is from a simple algebra statement. We know this statement is correct because it fits well with the Fibonacci series and the sequence obtained from a simple combination of successive Fibonacci numbers.

VII. The Newton Connection

Newton, known for centuries for his scientific work, was also a superb student of theology producing more material in theology than he did in science. This work remained stored in the libraries of Cambridge University, England and in private hands until 1936 when the collection was sold at auction. Cataloging of the collection and release of the manuscripts is ongoing. The most convenient place to access the collection as it unfolds is at the website maintained by the University of Sussex. [1]

Newton was proficient in Hebrew, Greek and Aramaic as well as Latin and English. He maintained that to fully understand a writer you had to read that writer in the language in which they wrote. This wide language proficiency was not unusual for scholars of his time. Interestingly, his successor as Lucasian Professor of Mathematics at the University of Cambridge, Henry Whiston, translated the complete works of Flavius Josephus (A. D. 37-100). Josephus, who many regard as the premier ancient historian, lived shortly after the time of Jesus.

It is impossible to cover all that Newton wrote concerning theology and prophecy, but there are several guiding principles that help in understanding his work in these areas and how they fit with this short account of Nature's Constants.

Newton's Views of Prophecy

1) Newton studied the ancient writings of the Jews and concluded, in accord with some of the ancient commentators on the Talmud, that the Jewish Messiah would make two appearances, or play two roles, one as a suffering Messiah and one as a conquering Messiah.

2) He believed that the lengths associated with earth distances and found elsewhere in nature were the key, along with descriptions of the Temples and the Ark of the Covenant, to understanding prophecy. These distances were related to time, possibly dates or

perhaps intervals of time between significant events. The timing of the second appearance of the Messiah was a subject he studied.

3) If, as Newton believed, the dimensions of the temples, especially Solomon's Temple, and dimensions in the earth were set by the Creator of the universe, then there should be a relationship between these dimensions and significant events in history. The reasoning on this was quite simple. If the same person were to build two different structures, you would expect some level of correlation between the designs. And if the builder were the Creator of the Universe, then perhaps the time of significant historical events would be coded into the structures. Further, since the temples were situated in Jerusalem, any distances to significant historical events should be measured from Jerusalem and time intervals or dates should be measured from significant events that occurred in Jerusalem or to the Jewish people.

His view of prophecy was that it was not exclusively to predict the future but to confirm the hand of the Creator through interpretation after certain historic events had taken place. Through the years there have been many who have predicted events and the time for those events based on their reading of prophecy. And most of them have been wrong about the timing and some about the events themselves. Newton spent considerable effort studying the book of Daniel and took special note of Daniel's admonition concerning how certain things should be "sealed up" until the "end of times" because they would not be understood except in the context of those times.

In relating distance there is the question of what length units to use, specifically nautical miles or statute miles. One hint of the appropriate units is that Babylon fell to the Medes late in the year of 539 BC, and it is 539 plus a fraction of a mile in statute miles from Babylon to Jerusalem. Note the juxtaposition of a significant event and a distance. It is 539 nautical miles from Jerusalem to Patmos where the Apostle John wrote the Book of Revelation, a book closely related to the Book of Daniel. The Book of Daniel was written in part during the Babylon captivity. Further, relating space to time works when distances from Jerusalem to events that occurred

42

in the BC era are measured in statute miles and distances to events that occurred in the AD era are measured in nautical miles. The modern state of Israel was established and recognized by other nations in the year 1948. This establishment had its roots when the British defeated the Ottoman Turks and took over administration of the area in 1917 followed by mandates by the British government supporting the establishment of a state of Israel. The distance from Jerusalem to London is 1948 nautical miles.

One year after the fall of Babylon to the Medes, in 538 BC the Jews were released. The establishment of Israel was in 1948 making 2486 years from the release from Babylon and the establishment of the modern state of Israel. At the time of the release of the Jews the Babylonians used a 360 day year, so multiplying the 2468 years by the current number of days in a year, 365.25, and then dividing by 360 days the time between release from Babylonian captivity to establishment of modern Israel in 1948 is 2520 years, the number written on the palace wall in on the eve of the fall of Babylon to the Medes. There is the problem of there being no zero year between AD and BC.

The city of Paris is 1799 nautical miles from Jerusalem. In the year 1799 Napoleon issued a proclamation of a Jewish state in Palestine. A few years later the plan was shelved primarily because advisors to Napoleon interpreted the scriptures to the effect that reestablishment of the Jewish state would usher in the times of the end of the world.

End Times Prediction

Before continuing on to Newton's specific prediction concerning the end times, one thing needs to be made clear here. The phrase "end of the world" is a popular one. Even those who use this phrase probably do not mean that the earth will disappear. A more accurate interpretation of the phrase is that "the end of the world" is the end of an era or is the time for a major change in world history. This unique historical event is often referred to by prophecy students as the beginning of the end times or simply the end times. And again the interpretation should be the end of an era times.

43

Students of Newton's papers were aware of his "end times" prediction many years before, but the prediction was popularized in 2003 by an article in the *London Daily Telegraph*. The calculation is based on the number 1260 which is one-half of 2520, the number written on the wall during Nebuchadnezzar's banquet on the evening Babylon fell, basically without armed conflict. And 2520 is one-half of 7!, 5040.

The calculation and logic are not that hard to follow. Newton assumed, as was the prevailing thinking at the time that the Messiah would return during a time when the world was ruled by a revised Roman Empire. Newton took the start of this empire rule to be 800 AD, the year of the coronation of Charlemagne as head of the revived Roman Empire. Adding 1260 years to the year 800 gives the year 2060 as the earliest date for the Messiah to return. In the 1660s this sounded good, but unfortunately for the calculation, the revised Roman Empire was dissolved in 1806, long after Newton's death possibly negating the logic associated with this calculation.

So the date may be wrong, but some aspects of the approach may be correct. Newton spent a lot of effort correlating the 1260 time period to the return of the Messiah beyond this simple calculation so we should not be quick to abandon this number. With the benefit of hindsight there is another interpretation of the starting time that fits with some other "end times" year calculations. The city of Rome was founded in 753 BC. David Flynn in the book *Temple at the Center of Time* suggests an allegorical view of the rebirth of the spiritual Rome as being mirrored about the time of the birth of Messiah making the spiritual year for the revised Roman Empire 753 AD.[2]

In the study of physical mirrors one of the simplest observations is that when you look in a mirror the object you see appears to be behind the mirror. If this approach is applied to time the analogous situation is as shown.

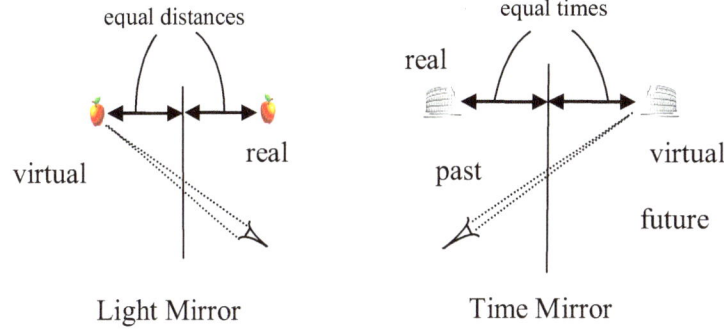

equal distances

equal times

real

virtual

real

past

virtual

future

Light Mirror

Time Mirror

Looking in the light mirror from the same side as the real image, the image appears behind the mirror.
Looking in the time mirror form the same side as the past event, the event appears on the other side of the time mirror or in the future.

Using Newton's logic to arrive at a time period of 1260 years, and the revised, or virtual, Roman Empire establishment at 753 AD, the earliest time for the return of the Messiah is 753 plus 1260 or 2013. Would the Creator of the Universe and the Controller of Events have coded this information into past history?

1. www.newtonproject.sussex.ac.uk.

2. *Temple at the Center of Time*, David Flynn, Anomatos Publishing, 2008

VIII. Conclusions

Numbers

Simple manipulations with numbers produce other numbers that occur in nature and have occurred historically. The factorial of 7 = 5040 was fascinating to the ancient Greeks. One-half and one-quarter of this number occur in history; notably 2520 is the number written on the wall at the fall of Babylon and 1260 appears in the dimensions of Avebury circle, and both numbers appear in the dimensions of the earth and moon.

The number π relates diameter to circumference in a circle. The number e comes from some very simple questions about number manipulation and is found in growth (of bacteria for example) and decay (radioactive decay for example) in many naturally occurring processes.

Measuring Systems

The nautical mile and the metric system, along with the number 360, form the basis of longitude and latitude navigation on the surface of the earth. And the correlation of longitudinal distance and time was a major scientific problem solved by the invention of accurate clocks. This time-distance relationship is most interesting.

The statute mile, the foot and the inch at first blush appear to be somewhat arbitrary, but on closer inspection they appear to be tied to constants appearing in nature.

If you are looking for correlation between numbers occurring in nature and construction, look no further than Avebury Circle, a very ancient structure.

Interesting Numbers

The definition of the golden ratio, either through algebra associated with length segments or from Fibonacci numbers, should convince even the skeptic that the Creator clearly enjoyed numbers. The preponderance of a Fibonacci number of petals on flowers is itself sufficient to show that the Creator really likes Fibonacci numbers. Or maybe He just likes flowers!

Newton is the primary researcher into the correlation of numbers and historical dates. He spent considerable effort correlating the numbers observed in nature with the distances between cities and the time between or dates of significant dates in history. His work concerning the "end times" and his prediction of the beginning of these "end times" is most important.

What about 2013?

There is one analysis that points to the year 2013 as being significant. Newton's calculation of the end times beginning after 2060 may be correct. He simply added 1260 years to what he considered the start of the Roman Empire in 800 AD. However, using the time image approach, and taking that the "virtual or spiritual" Roman Empire began in 753 BC, and then adding 1260 as Newton did, the beginning of the "end times" should start in 2013.

If there is a significant event that will occur in 2013, it should be a significant event associated with Jewish history. The best guess as to what that event is likely to be associated with events leading up to the second appearing of the Jewish Messiah.

It is interesting that Newton admonished his fellow Christians to not be misled by predictions as to the time of the Messiah's return but also to not be unaware of the eventuality of that event. He pointed out that some Christian's chided the Jewish community for not recognizing the first appearance of the Messiah while they themselves ignored his upcoming second appearance.

Appendix A The Longer Earth Day

The historical record is clear that at one time the year, the time for one complete orbit of the earth around the sun, consisted of 360 days. Now, however, it is 365.25 days. As explained so well by Velikofsky, there was in the past a major and sudden shift in the orientation of the rotational axis of the earth. This change in rotational orientation was in the neighborhood of 90 degrees. What are now arctic regions were once lush forests populated by large mammals.

Virtually every culture records a great flood in past times. Putting the change in the length of the day together with a great flood, Velikofsky presents an argument that these two changes, the change in rotation orientation and the great flood, were produced by a massive visitor passing close by the earth, dumping a large amount of ice, much of which was turned into water, and affecting the rotational orientation of the earth and the length of the year.

The earth orbiting about the sun is subject to the gravitational attraction of the sun. This gravitational force exerted on the earth by the sun follows the Universal Law of Gravitation which states that the force exerted on the earth by the sun is proportional to a universal constant and the product of the masses of the two bodies involved, the sun and the earth, divided by the square of the distance between their centers. In mathematical form this is

$$F_g = G \frac{M_e M_s}{R_{e-s}^2}$$

The universal gravitational constant, G, is well established by experiment, and the masses of the earth and sun are well known as is the average distance between their centers. Other bodies contribute to the total gravitational force on the earth, and the orbit of the earth around the sun is elliptical and not really a circle, but the main gravitational force on the earth is this one.

This gravitational force supplies the centripetal (center-directed) force that makes the earth move in a circle. The gravitational force on the earth due to the sun points in the direction from the earth to the sun. This at first sounds a bit odd. Here is the earth moving in a circle with the direction of the velocity tangent to the circle at every point and the force at right angles to this velocity directed toward the center of the circle.

Before you begin to worry that the system looks as if it should slow down with the earth spiraling into the sun, look at a simple experiment that we can carry out with our minds and a little experience. Whirl a mass attached to the end of a string in a horizontal circle. The only force applied is through the string and is directed toward the center of the circle. This force depends on the mass, the velocity of the mass moving in the circle and the radius of the circle according to

$$F_c = \frac{mv^2}{r}$$

where m represents the mass, v the velocity and r the radius of the circle. Spin the mass faster and there is more force on the string. Increase the radius and the force is decreased. Experiments similar to this are performed in most elementary physics courses. Students whirl a rubber stopper on the end of a string with a glass rod. The string is from the mass over to the glass rod and then down to a spring scale that measures force. Knowing the radius and timing the rotations provides the velocity, radius and with a known mass the force can be calculated and compared to the measured force.

In the experiment the mass is whirled about, say 10 times taking the time for the ten rotations. This provides the average time for one rotation, and knowing the radius the distance around the circle divided by the average time for one rotation, gives the velocity. Then with the known radius and mass the force can be calculated. This calculated force is compared to the force as measured on the force meter.

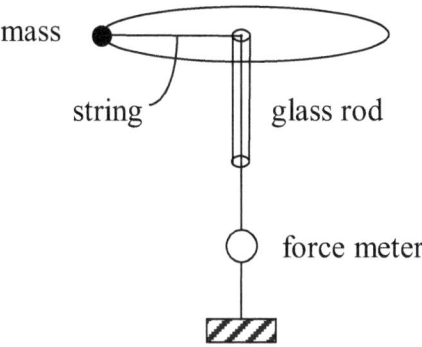

mass

string

glass rod

force meter

For the earth-sun system the gravitational force provides this center directed force that keeps the earth rotating in a circle about the sun. The balance statement is from equating these two forces.

$$G\frac{M_e M_s}{R_{e-s}^2} = \frac{M_e V^2}{R_{e-s}}$$

If the mass of the earth is increased due to an ice dump, this has no effect on the balance because the mass of the earth is on both sides of the equation. If the mass of the sun remains constant and the velocity decreases, the only thing left in the equation that can change is the earth-sun distance. Reducing the equilibrium statement by crossing off common terms gives a better feel for how this might work.

$$G\frac{M_s}{R_{e-s}} = V^2$$

If the velocity of the earth is slowed going from 360 days per year to 365.25 days, then the change is 5.25 in 360. This fraction works out to 0.0146 or about 1.5 %. The new balance equation then would read

51

$$G\frac{M_s}{R_{e-s}} = \left[V(1-0.015)\right]^2 = \left[(0.985)V\right]^2 = (0.970)V^2$$

If G and M_{e-s} remain constant then this equation can be written as

$$GM_s = R_{e-s}(0.970)V^2$$

And for this situation the earth-sun distance would have to increase by a factor of 1.03 or roughly 3%.

It is hard to carry this reasoning much further. The first consideration would be how this increase in average distance from the sun might affect the temperature of the planet. Increasing the mass of the earth, especially with water and ice, and reorienting the axis of rotation conceivably could dramatically change the absorption of energy delivered by the sun. Furthermore, the massive amounts of water deposited might affect the composition of the atmosphere thus affecting the amount of radiation that reached the planet. To carry this scenario much further would entail much speculation. There are just too many variables and too many unknowns.

www.ingramcontent.com/pod-product-compliance
Lightning Source LLC
Chambersburg PA
CBHW041110180526
45172CB00001B/191